THE POETRY OF MAGNESIUM

The Poetry of Magnesium

Walter the Educator™

SKB

Silent King Books a WhichHead Imprint

Copyright © 2023 by Walter the Educator™

All rights reserved. No part of this book may be reproduced in any manner whatsoever without written permission except in the case of brief quotations embodied in critical articles and reviews.

First Printing, 2023

Disclaimer
This book is a literary work; poems are not about specific persons, locations, situations, and/or circumstances unless mentioned in a historical context. This book is for entertainment and informational purposes only. The author and publisher offer this information without warranties expressed or implied. No matter the grounds, neither the author nor the publisher will be accountable for any losses, injuries, or other damages caused by the reader's use of this book. The use of this book acknowledges an understanding and acceptance of this disclaimer.

"Earning a degree in chemistry changed my life!"
- Walter the Educator

dedicated to all the chemistry lovers, like myself, across the world

CONTENTS

Dedication	v
Why I Created A Book?	1
One - Symbol Of Strength	2
Two - Precious Element	4
Three - Depths Of The Sea	6
Four - Magnesium Shines	8
Five - Metal Divine	10
Six - Secrets You Keep	12
Seven - Element Of Grace	13
Eight - Cosmic Gem	15
Nine - Perfect Harmony	17
Ten - Divine Source	19
Eleven - Beacon In The Dark	21
Twelve - Radiant And Pure	23

Thirteen - Nature's Embrace 24

Fourteen - Tapestry Of Wonder 26

Fifteen - Magnesium, The Element 28

Sixteen - Countless Ways 30

Seventeen - You Are Divine 32

Eighteen - Near And Far 34

Nineteen - The Conductor 36

Twenty - Vital Contribution 38

Twenty-One - Fierce And Subtle 40

Twenty-Two - Oh Magnesium 42

Twenty-Three - Vibrant Hue 44

Twenty-Four - Science And Art 46

Twenty-Five - Crucial Part 47

Twenty-Six - Dazzling Display 48

Twenty-Seven - DNA, Muscles, Photosynthesis, And More 50

Twenty-Eight - Across The Land 52

Twenty-Nine - Beauty, Passion, And Love . . 54

Thirty - Oh, Magnesium 56

Thirty-One - Magnificent Jewel 58

Thirty-Two - Guardian Of Life	60
Thirty-Three - Fireworks' Glow	62
Thirty-Four - Without You	64
Thirty-Five - Power So Potent	65
Thirty-Six - Versatile Ways	67
About The Author	69

WHY I CREATED A BOOK?

Creating a poetry book about the chemical element of Magnesium can be a unique and intriguing concept. Magnesium, a versatile and abundant element, possesses various properties and applications that can serve as metaphors and symbols in poetry. Exploring the physical and chemical characteristics of Magnesium, such as its brightness, lightness, and reactivity, inspires me to delve into themes of illumination, transformation, and the human condition. By intertwining scientific knowledge with artistic expression, this poetry book can offer a new perspective on the world and ignite curiosity in readers about the wonders of science and poetry.

ONE

SYMBOL OF STRENGTH

In the realm of elements, a shining star,
Magnesium, we marvel at what you are.
With atomic number twelve, you hold your place,
A metal, strong and full of grace.

Your name derived from the Greek, a tale so old,
Magnesia, a land where your secrets unfold.
In nature's realm, you're found abundantly,
A gift bestowed upon us, so brilliantly.

Your gleaming presence, a sight to behold,
A silver-white metal, radiant and bold.
With a fiery passion, you burn so bright,
Igniting the darkness with your dazzling light.

From stars to fireworks, your sparks ignite,
A spectacle of brilliance, a celestial sight.
In alloys, you strengthen, and corrosion you fight,
A protector, a guardian, shining with might.

In our bodies, you play a vital role,
A mineral essential, for mind and soul.
From bones to nerves, you support and heal,
Magnesium, a gift that life can't conceal.

So, let us celebrate your wondrous might,
Magnesium, a beacon, shining so bright.
In the world of elements, you take your place,
A symbol of strength, elegance, and grace.

TWO

PRECIOUS ELEMENT

In the realm of elements, a shining star,
Magnesium, strong and graceful from afar.
Its name derived from Magnesia's ancient land,
Where its abundance in nature was first found.

Igniting brilliance with a fiery hue,
Magnesium, oh how we marvel at you.
With sparks that dance and flames that ignite,
You illuminate the darkness with your light.

Within our bodies, you hold a vital role,
A catalyst for reactions, a key to the whole.
From our bones to our muscles, you lend your might,
Keeping us strong and flexible, day and night.

Oh, Magnesium, symbol of strength and grace,
In the world of elements, you hold your place.
A dazzling presence, both fierce and serene,
A testament to the wonders that nature has seen.

So let us celebrate this precious element,
Its power and beauty, so evident.
Magnesium, we salute you with awe,
As you continue to inspire and ignite us all.

THREE

DEPTHS OF THE SEA

In the realm of elements, there lies Magnesium,
A force of nature, a marvel to behold.
With strength and grace, it shines with elegance,
A beacon of power, a story yet untold.
    Like a warrior, it stands tall and unyielding,
A guardian of life, a symbol of might.
Its atoms dance in harmony, revealing
A symphony of power, an eternal light.
    From starlit skies to the depths of the sea,
Magnesium weaves its magic, a cosmic thread.
In every leaf and flower, its touch sets free
The colors of creation, the beauty widespread.
    Oh Magnesium, you ignite the night,
With sparks of brilliance, you light the way.
A catalyst of dreams, a guiding light,
You inspire us to reach, to never sway.

In our bodies, you flow, a vital force,
Nurturing our bones, our muscles, our soul.
With every beat of our hearts, we endorse
Your presence, your purpose, our lives made whole.
Magnesium, oh Magnesium, we sing your praise,
For the strength you lend, for the dreams you raise.
A symbol of power, a touch of grace,
You inspire us to shine, to conquer space.

FOUR

MAGNESIUM SHINES

In nature's embrace, a radiant light,
Magnesium shines, a spectacle so bright.
A dancer of fire, it burns in the night,
With a brilliance that captivates, pure and white.

Within the Earth's core, it quietly resides,
A treasure untapped, where its power hides.
A metal so strong, it defies the tides,
A force that nature skillfully abides.

In our bodies, Magnesium plays a role,
A vital element, it keeps us whole.
From bones to muscles, it takes control,
Supporting life's functions, body and soul.

From stars in the skies to the depths of the sea,
Magnesium's presence, a wonder to see.
A symbol of power and beauty set free,
Inspiring awe and curiosity.

Oh, Magnesium, with your mystical grace,
You illuminate the world, leaving no trace.
A symbol of strength, in every embrace,
Forever enchanting, in time and in space.

FIVE

METAL DIVINE

In the realm of elements, a star does rise,
A metal of power, Magnesium's prize.
With atomic fire, its brilliance resounds,
A luminescent beauty, nature astounds.

From the depths of Earth, it emerges bright,
Igniting flames, a celestial light.
In chlorophyll's embrace, it dances free,
Bestowing life's vigor to every tree.

In our bodies, it dwells, a vital force,
Enriching bones, its presence endorses.
Muscles contract, nerves fire with grace,
Magnesium's touch, a rhythmic embrace.

Through the cosmos it travels, a cosmic ray,
In supernovas, it finds its way.
A fusion of stars, an alchemical blend,
Magnesium, the luminary, till the end.

So let us celebrate this metal divine,
A beacon of radiance, forever to shine.
Magnesium, the element that inspires,
Unleashing our spirits, setting our souls on fire.

SIX

SECRETS YOU KEEP

In nature's embrace, Magnesium thrives,
A radiant element, where brilliance resides.
From earth's deep womb, it emerges with might,
A treasure of fire, igniting the night.

In flames it dances, a shimmering glow,
A spectacle of light, captivating the show.
With sparks and sparks, it paints the skies,
A cosmic display, to mesmerize.

In our bodies, it plays a vital role,
Building bones and muscles, making us whole.
A partner to enzymes, it aids in their task,
Catalyzing reactions, no questions to ask.

Magnesium, oh radiant star,
In nature and within us, you travel far.
From soils to stars, your essence does seep,
A glowing reminder, of the secrets you keep.

SEVEN

ELEMENT OF GRACE

In the realm of elements, Magnesium stands tall,
A shining star, beloved by all.
With atomic number twelve, it's no pretender,
And its properties, oh, they're truly splendid.

Silver-white and lustrous, it catches the eye,
A metal so versatile, it can't be denied.
Its strength and lightness make it a treasure,
In industries, it finds wide-scale use and pleasure.

In the fires of stars, it was born,
A gift from the cosmos, now widely worn.
From flares to fireworks, it ignites the night,
In a dazzling display, it takes its flight.

But Magnesium's magic does not end there,
For in our bodies, it plays a crucial affair.
It helps our muscles, nerves, and bones,
Regulating their functions and making them strong.

From the beating of our hearts to the thoughts in our minds,
Magnesium supports us, in ways we can't find.
It calms our worries, eases our stress,
A mineral so vital, we must confess.
So let us cherish this element of grace,
Magnesium, with its radiant embrace.
In science and life, it holds its sway,
A testament to nature's wondrous display.

EIGHT

COSMIC GEM

In nature's realm, a starry glow,
A secret tale that few may know,
A metal rare, with brilliance bright,
Magnesium, a cosmic light.

From ancient seas, it did arise,
A gift bestowed from azure skies,
Its flame, a dance of pure delight,
Magnesium, a radiant sight.

Within our bones, its presence lies,
A sturdy frame, it fortifies,
A guardian, steadfast and strong,
Magnesium, where we belong.

A catalyst, it plays its part,
In every beat of our own heart,
Enzymes dance, the rhythm flows,
Magnesium, life's symphony grows.

    From chlorophyll to vibrant green,
A messenger, unseen, unseen,
In every leaf, a vital role,
Magnesium, nature's precious soul.
    So let us honor, let us praise,
This element that lights our ways,
Magnesium, a cosmic gem,
Forever shining, within us, them.

NINE

PERFECT HARMONY

In the heart of stars, a brilliance untold,
A cosmic dance of fire, untamed and bold.
Magnesium, the element that ignites the flame,
A radiant force, bearing a celestial name.

In nature's embrace, it weaves its design,
From earth's rich soil to oceans' brine.
A guardian of life, in chlorophyll it resides,
Painting leaves green, a symphony of tides.

In enzymes, it dances, a catalyst of might,
Unleashing reactions, a wondrous sight.
A conductor of change, it sparks the way,
In the alchemy of life, it has a say.

Within us it dwells, a silent guide,
Nurturing bones, where strength does reside.
Muscles and nerves, it gently sustains,
A guardian of balance, where harmony reigns.

Yet beyond our grasp, it reaches the stars,
Bathing the cosmos in its radiant scars.
A celestial wanderer, it enchants and inspires,
A cosmic dancer, setting hearts afire.

Magnesium, mysterious and profound,
In nature's tapestry, a treasure found.
From earth to sky, its presence we see,
A symphony of elements, in perfect harmony.

TEN

DIVINE SOURCE

In the depths of the earth, a treasure concealed,
A shining element, Magnesium revealed.
With a gleam that rivals the sun's golden rays,
Its power and beauty, forever amaze.

It dances with fire, a spark in the night,
Igniting the darkness, a celestial light.
In our bodies, it dances, an essential role,
Nourishing our being, it makes us whole.

From the beating heart to the bones so strong,
Magnesium's presence, never does it wrong.
It strengthens our muscles, with every stride,
And keeps our energy flowing, far and wide.

In the embrace of nature, it's found in great measure,
In the leaves of the trees, bringing vibrant pleasure.

In the ocean's embrace, where life does thrive,
Magnesium, the element, keeps our souls alive.
   Oh, Magnesium, element of grace,
In every molecule, you leave your trace.
A symbol of power and life's vital force,
Magnesium, you are nature's divine source.

ELEVEN

BEACON IN THE DARK

In the realm of elements, Magnesium shines,
A radiant light that ignites the night.
Its brilliance, a spectacle divine,
A cosmic spark, forever burning bright.

Within the earth, it dwells, a noble guest,
A catalyst for life, a force untamed.
In nature's realm, it weaves its steady quest,
Enchanting all with beauty unconstrained.

From stars above to oceans' depths below,
Magnesium dances with the cosmic tide.
It molds the world, a sculptor in its flow,
A conductor of change, no bounds to hide.

In bones and sinews, it finds its embrace,
A steadfast ally, strong and true.

It lends its strength, a touch of grace,
A foundation for life, both old and new.
    So let us honor this element of might,
Magnesium, the beacon in the dark.
In body and soul, it guides us through the night,
A celestial light, leaving its eternal mark.

TWELVE

RADIANT AND PURE

In the realm of elements, radiant and pure,
A shimmering light that will endure.
Magnesium, the mighty in its own right,
An element that shines with celestial might.

Within our bodies, it plays a vital role,
A catalyst for life, it takes control.
From bones to muscles, it lends its aid,
In every heartbeat, its presence is laid.

In nature's tapestry, it weaves its thread,
A symphony of growth, where life is fed.
From plants to creatures, it fosters the dance,
A conductor of change, a cosmic chance.

And so, Magnesium, we honor your might,
A beacon of strength, forever alight.
In our bodies and in nature's embrace,
You leave an eternal mark, a celestial grace.

THIRTEEN

NATURE'S EMBRACE

In the realm of life's alchemy,
A star-born element, bold and free,
Magnesium, your essence divine,
In every atom, you brightly shine.

Within our bodies, a steadfast force,
You weave your magic, a vital source,
Muscles contract, nerves ignite,
With every heartbeat, you bring us light.

A catalyst for life's symphony,
You dance with atoms, harmoniously,
From bones to cells, you play your part,
Nurturing life with your gentle art.

In nature's tapestry, you paint,
A golden hue, a celestial saint,
From fiery sunsets to moonlit nights,
Your radiance fills our earthly sights.

Magnesium, your power untold,
In secrets of stars, your story unfolds,
With cosmic grace, you inspire and enchant,
A celestial jewel, forever gallant.
So let us celebrate your splendid reign,
Magnesium, element of strength, not in vain,
For in our bodies and nature's embrace,
Your brilliance shines, leaving a lasting trace.

FOURTEEN

TAPESTRY OF WONDER

In the alchemy of life, where secrets reside,
Magnesium, a star, does quietly abide.
A luminescent metal, so brilliantly bright,
It dances with flames, igniting the night.

With a fiery spirit, it sparks and inspires,
Transforming the ordinary to extraordinary desires.
In nature's embrace, it weaves its spell,
From the depths of the earth to the heavens, it dwells.

The chlorophyll's guardian, it breathes life anew,
In leaves of green, a vibrant, verdant hue.
A conductor of energy, it powers the sun,
Through photosynthesis, the cycle is spun.

In the human vessel, it plays a vital role,
Nourishing the body, a nutrient for the soul.
It strengthens the bones, a guardian so strong,
A protector of muscles, where resilience belongs.

Magnesium, a catalyst, a conductor of change,
In biochemical reactions, it rearranges.
From heartbeat to nerve impulses, it's at the core,
A guardian of balance, forevermore.

In the alchemy of life, Magnesium weaves,
A tapestry of wonder, where magic conceives.
A force of nature, both gentle and grand,
A symphony of elements, held in its hand.

FIFTEEN

MAGNESIUM, THE ELEMENT

In realms of fire and luminous grace,
Magnesium, the conductor, takes its place.
A catalyst for change, a force untamed,
In every atom, its power contained.

In the human frame, a guardian of balance,
Magnesium dances, a vital alliance.
Enthroned in bones, it stands tall and true,
Supporting life's structures, through and through.

In muscles, it whispers, a silent command,
Contracting and relaxing, a symphony on demand.
Nerves it energizes, impulses it ignites,
Guiding the body through day and night.

In nature's embrace, Magnesium weaves,
A tapestry of wonders, where magic conceives.

In chlorophyll's embrace, it captures the sun,
Transforming light into life, a battle won.
   From stars to soil, it travels unseen,
A cosmic traveler, a messenger keen.
In fireworks' burst, it dazzles the sky,
A pyrotechnic wonder, captivating the eye.
   Oh, Magnesium, conductor of energy's flow,
A luminary force, forever aglow.
In every heartbeat, a rhythm it keeps,
A symphony of life, where harmony leaps.
   So let us celebrate this element divine,
For its presence, a gift, a treasure to find.
In the world of science, an enigma profound,
Magnesium, the element, forever renowned.

SIXTEEN

COUNTLESS WAYS

Magnesium, catalyst of life,
A force untamed, a flame of might.
In the depths of Earth, you reside,
A shimmering metal, strong and bright.

With burning brilliance, you ignite,
A dance of atoms, pure and white.
From stars to soil, you gracefully flow,
Unleashing energy, aglow.

In every cell, you play your part,
A vital element, from the start.
Enzymes and proteins, you bring to life,
With every beat, with every stride.

Magnesium, oh radiant one,
A conductor of life's symphony,
You orchestrate the melody,
In every rhythm, in every key.

    From chlorophyll in leaves so green,
To DNA, a wondrous scene.
You guide the dance of life's ballet,
A silent hero, in every way.
    So let us honor your brilliance,
Your power, your divine essence.
Magnesium, we sing your praise,
A bringer of light, in countless ways.

SEVENTEEN

YOU ARE DIVINE

Magnesium, you catalyst of life,
A helper in the dance of time,
A spark to start the fire's blaze,
And guide the way through endless space.
In nature, you're a common sight,
In chlorophyll, your power's might,
You help the plants to grow and thrive,
And keep the balance, help them survive.
Oh Magnesium, you mighty force,
In humans, you're the source of course,
A strength to keep the heart in check,
And help the bones to heal and connect.
Through energy, you conduct the charge,
And in biochemistry, you play a large,
A role in reactions, a key to life,
Magnesium, you make the world so bright.

So let us celebrate your worth,
Your contributions to this earth,
Magnesium, you are divine,
A treasure that we'll always find.

EIGHTEEN

NEAR AND FAR

In the realm of energy, it does reside,
A conductor of power, with a radiant stride.
Magnesium, the element, so pure and bright,
Igniting the flame, illuminating the night.

In the depths of the body, it weaves its spell,
Enzymes and proteins, it knows them well.
From muscle contractions to nerve transmissions,
Magnesium dances, with graceful precision.

A guardian of bones, it stands tall and strong,
Supporting the structure, where life belongs.
It whispers to cells, in a language so fine,
Guiding the dance, the rhythm, the line.

In DNA's spiral, it finds its place,
Connecting the strands, with elegance and grace.
A messenger of life, it carries the code,
Unraveling the secrets, that nature bestowed.

Oh, Magnesium, you dazzle in the dark,
In fireworks, you leave your fiery mark.
A burst of brilliance, a vibrant display,
A testament to your presence, in nature's array.

So let us celebrate, this element profound,
For its role in life's dance, we are forever bound.
Magnesium, the conductor, the shimmering star,
A symbol of energy, both near and far.

NINETEEN

THE CONDUCTOR

In the realm of energy, a conductor supreme,
Magnesium, a metal, with a vibrant gleam.
Within the human body, it finds its place,
A vital element, in every single trace.

Enzymes it activates, a catalyst so grand,
From ATP synthesis to DNA strand.
In muscles it dances, contracting with might,
Aiding in movements, bringing strength to the fight.

Photosynthesis owes its brilliance to thee,
For Chlorophyll's creation, you set it free.
In leaves of green, the sun's rays you embrace,
Transforming light to energy, with grace.

Oh, Magnesium, in fireworks you shine,
With sparks and colors, you make the night divine.
A spectacle of brilliance, a dazzling display,
Your magic in the sky, we celebrate and sway.

So, let us give homage to this element so dear,
Magnesium, the conductor, we hold you near.
From biochemical reactions to life's very core,
You guide the dance of energy, forevermore.

TWENTY

VITAL CONTRIBUTION

In the realm of nature's grace,
A luminary element takes its place.
Magnesium, a star in the periodic sky,
With secrets hidden, waiting to comply.

In chlorophyll's green embrace,
Photosynthesis finds its solace.
With sun's rays, a dance begins,
Where Magnesium's magic never dims.

Bones and muscles, strong and true,
Magnesium, we owe it all to you.
With every step and every stride,
Strength and resilience, you provide.

A catalyst in biochemical reactions,
Magnesium, the heart of chemical attractions.
Enzymes awaken, pathways unfold,
Life's intricate web, you gently mold.

And now, a different tale to tell,
Where energy conducts, and wonders dwell.
Fireworks ignite, in colors so grand,
Magnesium's presence, a fiery brand.

Explosions in the night, a dazzling array,
Magnesium, the conductor of this display.
Sparkling lights, a symphony of flame,
A testament to your vibrant, fiery name.

DNA's blueprint, the code of life,
Magnesium, you guide the dance so rife.
A choreographer of genetic expression,
Unraveling the mysteries of life's progression.

In muscle contractions, you play a part,
Magnesium, the conductor of the beating heart.
With every pulse, you orchestrate,
The rhythm of life, in perfect state.

And as the leaves turn toward the sun,
Photosynthesis, your work is done.
Magnesium, the spark of life's creation,
We marvel at your vital contribution.

TWENTY-ONE

FIERCE AND SUBTLE

In nature's realm, a starlit gleam,
Magnesium, the element supreme.
With brilliance that the eye adores,
It dances through biochemical shores.

In every cell, its presence found,
A vital link, profound and sound.
It sparks the engines of life's design,
Catalyzing reactions, oh so fine.

From DNA's spiral, a code untold,
Magnesium's touch, a story unfolds.
It weaves the strands, a delicate chore,
Unraveling secrets, forevermore.

In muscles strong, it takes its place,
Contracting fibers with steadfast grace.
Enabling movement, both swift and true,
Magnesium, the conductor through and through.

And in the night's sky, a dazzling sight,
Fireworks ignite, with colors bright.
It's Magnesium's spark that paints the air,
A symphony of light, beyond compare.

So let us celebrate this wondrous metal,
With its power untamed, both fierce and subtle.
In biochemistry's realm, a star it shines,
Magnesium, the treasure of our times.

TWENTY-TWO

OH MAGNESIUM

In nature's secret alchemy, behold,
A shimmering element, pure and bold.
Magnesium, the star of Earth's domain,
Ignites the fervor in life's vibrant chain.

In biochemistry's intricate dance,
Magnesium weaves its spell, perchance.
A catalyst of enzymes, it plays its part,
Unleashing energy from every heart.

In DNA's helix, a sacred script,
Magnesium's presence, tightly gripped.
A stabilizing force, it binds the strands,
A molecular symphony, elegantly planned.

Muscle contractions, a rhythmic beat,
Magnesium's touch, a pulse so sweet.
From twitch to flex, it fuels the fire,
Empowering movement, lifting us higher.

In photosynthesis' radiant glow,
Magnesium orchestrates the show.
Absorbing light, it channels the energy,
Transforming sunlight into life's synergy.

And in the night sky's dazzling array,
Fireworks burst, a grand display.
Magnesium's brilliance, a starry delight,
Painting the heavens in colors so bright.

Oh Magnesium, element divine,
A symbol of life's grand design.
From Earth to cosmos, your power unfurled,
A testament to the wonders of this world.

TWENTY-THREE

VIBRANT HUE

In nature's realm, where wonders lie,
A metal gleams 'neath azure sky.
Magnesium, its name so grand,
A marvel crafted by nature's hand.

Within biochemistry's intricate dance,
Magnesium weaves its magical trance.
Enzymes it activates with gentle touch,
A catalyst, igniting life's clutch.

In fireworks' majestic display,
Magnesium takes center stage, they say.
With sparks and flames, it dazzles the night,
A symphony of colors, pure delight.

Oh, Magnesium, element divine,
In DNA's code, you intertwine.
A backbone strong, a structure true,
Guiding life's secrets, old and new.

Muscle fibers, they contract and flex,
Magnesium, the conductor, directs.
In every beat, in every move,
Its presence fills the rhythmic groove.

   And in the green of nature's grace,
Photosynthesis finds its rightful place.
Magnesium's chlorophyll, a vibrant hue,
Captures sunlight, life to renew.

   Magnesium, oh element rare,
With each new poem, we lay you bare.
Your beauty and significance, forever entwined,
In science and nature, an eternal bind.

TWENTY-FOUR

SCIENCE AND ART

In the realm of elements, Magnesium shines,
A marvel of nature, a treasure divine.
In DNA's code, its presence is found,
A building block of life, profound.

Muscles contract with its guiding force,
Strength and movement, a powerful source.
Photosynthesis dances in its embrace,
Capturing light, sustaining life's grace.

Oh, Magnesium, you light up the night,
In fireworks' bursts, a colorful sight.
With hues so vibrant, you paint the sky,
A spectacle that makes hearts soar high.

From genetic expression to fiery delight,
Magnesium weaves wonders, day and night.
In every facet of life, you play your part,
A catalyst of beauty, in science and art.

TWENTY-FIVE

CRUCIAL PART

Of all the elements in the periodic table,
Magnesium stands out as unique and able.
It helps our muscles contract and expand,
And plays a role in photosynthesis so grand.
    In DNA, it's a crucial part,
Helping to keep our genetic code smart.
And in fireworks, it shines so bright,
A dazzling display of color and light.
    Magnesium, oh Magnesium, what a wonder you are,
A versatile element, you shine like a star.
From the depths of the earth to the depths of space,
Your presence is felt in every place.

TWENTY-SIX

DAZZLING DISPLAY

In the realm of elements, Magnesium shines bright,
A star in the periodic table, a captivating sight.
With atomic number twelve, it stands with grace,
Unveiling its secrets, leaving none in trace.

In the realms of life, Magnesium does roam,
Playing crucial roles, making a vibrant home.
Within our DNA, it weaves its magical spell,
A silent architect, where life's stories dwell.

In muscle contractions, it takes center stage,
Fueling the movements, like a symphony on a page.
From a twitch to a leap, from a bend to a run,
Magnesium dances, ensuring the work is done.

In photosynthesis, it's an essential player,
Absorbing light energy, a true nature slayer.
Chlorophyll's partner, in the greenest of hues,
Magnesium's presence, a gift we can't refuse.

And in the night sky, it paints a dazzling display,
Fireworks explode, as if stars have found their way.
Magnesium's brilliance, a spectacle so grand,
Bursting with colors, like an artist's skilled hand.

So let us celebrate this element divine,
Magnesium, oh Magnesium, forever shall you shine.
In the depths of science, and in nature's embrace,
Your contributions, forever leave a trace.

TWENTY-SEVEN

DNA, MUSCLES, PHOTOSYNTHESIS, AND MORE

In the realm of elements, a star does gleam,
With brilliance that enchants, a radiant beam.
Magnesium, the name that graces the stage,
A versatile element, a marvel to engage.

In the dance of life, where DNA weaves,
Magnesium plays a role, a bond it achieves.
For in the helix of genes, it stands tall,
A vital part, within the DNA's thrall.

Muscles, they contract, with strength and might,
Magnesium, the conductor, orchestrates the fight.
From twitch to leap, from flex to bend,
It fuels the movements, without an end.

In the realm of nature, where sunlight weaves,

Photosynthesis unfolds, as Magnesium conceives.
Within the chlorophyll's embrace, it resides,
Absorbing light, where life abides.

And when the night sky bursts in colorful bloom,
Fireworks ignite, dispelling the gloom.
Magnesium, the spark, the dazzling show,
A spectacle of light, a radiant glow.

So let us celebrate this element divine,
Magnesium, a treasure, forever will shine.
Through DNA, muscles, photosynthesis, and more,
Its presence we cherish, forever adore.

TWENTY-EIGHT

ACROSS THE LAND

In the realm of atoms, a star does reside,
A shining element, with a glow so wide.
Magnesium, the name that we adore,
A metal so pure, it's worth to explore.

In muscles, it dances, with grace and might,
Contracting and relaxing, both day and night.
It fuels our movements, with every stride,
A conductor of strength, deep down inside.

In the realm of nature, a marvel it shows,
In green leaves and plants, where life freely flows.
Photosynthesis, a magical dance,
Magnesium's presence, a vital advance.

And let's not forget, the colorful skies,
Where fireworks burst, with radiant highs.
Magnesium's sparks, a spectacle grand,
Illuminating the heavens, across the land.

So, let us celebrate, this element so rare,
With its wonders and mysteries, beyond compare.
From DNA's structure, to muscle's might,
Magnesium shines, in the day and night.

TWENTY-NINE

BEAUTY, PASSION, AND LOVE

In the realm of elements, let us gaze,
Upon the beauty of Magnesium's blaze.
A metal of wonder, so bright and pure,
Its secrets allure, forever endure.

In muscles, it dances, a vital role,
Contraction and movement, it does control.
From head to toe, in every limb,
Magnesium's touch makes muscles trim.

Photosynthesis, a magical feat,
Where sunlight and chlorophyll meet.
Magnesium, a partner in this grand dance,
Nurturing plants with its elegant stance.

And in the night sky, a dazzling sight,
Fireworks explode, colors ignite.

Magnesium's spark, a vibrant display,
Filling the air, in a magnificent array.
   So let us celebrate, Magnesium's might,
A versatile element, shining bright.
From muscles to plants, and the stars above,
Its presence brings beauty, passion, and love.

THIRTY

OH, MAGNESIUM

In the realm of elements, a star is born,
A gift to nature, Magnesium, adorn.
Within our bodies, its presence prevails,
A catalyst for life, where strength prevails.

Muscle contractions, a symphony of might,
Magnesium's touch, ignites the fight.
From twitch to flex, it sparks the flame,
The dance of motion, a majestic game.

In photosynthesis, it plays its role,
A pigment of green, a vital soul.
Chlorophyll's partner, so pure and bright,
Capturing sunlight, bringing life's light.

And in the night sky, a spectacle, grand,
Magnesium's magic, in fireworks, expand.
A burst of color, a shower of delight,
A symphony of sparks, painting the night.

Oh, Magnesium, element so divine,
Your presence, a blessing, through space and time.
In muscle and leaf, in the night's embrace,
Your essence, a marvel, a touch of grace.

THIRTY-ONE

MAGNIFICENT JEWEL

In the realm of elements, Magnesium shines,
A wondrous metal, with powers so fine.
Within our bodies, it plays a key role,
In muscles, it helps us stay strong and whole.

When we move, it aids in contraction,
Aiding our bodies with every action.
From head to toe, it helps us flex,
Without Magnesium, we'd be motionless wrecks.

In nature's workshop, it's a vital tool,
For photosynthesis, a magnificent jewel.
Within the chlorophyll, it stands tall,
Converting sunlight into energy, overall.

And oh, the beauty it brings to the night,
In fireworks, it dazzles with colorful light.
With a burst of brilliance, it lights up the sky,
Magnesium's spark, it makes us sigh.

So let us celebrate this element's might,
From muscles to fireworks, it shines so bright.
Magnesium, a hero in its own right,
A treasure that fills our world with delight.

THIRTY-TWO

GUARDIAN OF LIFE

In the heart of nature, Magnesium resides,
A shimmering presence, where beauty abides.
Its atomic dance, a celestial display,
Igniting the stars in a cosmic ballet.

From the depths of the Earth, it rises with might,
A metal of wonders, casting its light.
In chlorophyll's embrace, it weaves a grand plot,
Empowering life with a photosynthetic shot.

Behold the leaves, a verdant symphony,
Magnesium's touch, a key to harmony.
With each breath they take, a dance of delight,
Transforming sunlight into energy so bright.

In muscles, it pulses, a powerful force,
Fueling the movements, a wild, untamed source.
From the twitch of a finger to the leap of a bound,
Magnesium's magic, in every muscle found.

So let us marvel at this element rare,
A guardian of life, beyond compare.
From the cosmos to cells, it weaves its grand tale,
Magnesium, the star of nature's wondrous trail.

THIRTY-THREE

FIREWORKS' GLOW

In the realm of fire and light,
A miracle element shines so bright.
Its name is Magnesium, a wondrous sight,
Igniting the heavens with colors so tight.

In the darkest of nights, it takes flight,
Explosions of brilliance, a dazzling height.
In fireworks it dances, a vibrant sprite,
Painting the sky with its radiant might.

But Magnesium's wonders extend far beyond,
For in nature's grand design, it's fondly fond.
In muscle movement, it's a hero profound,
Contracting fibers, strength unbound.

Photosynthesis, a magical affair,
Magnesium's presence, it cannot spare.
From the chlorophyll green to the oxygen's share,
It breathes life into trees with utmost care.

So let us marvel at Magnesium's grace,
In fireworks' glow and nature's embrace.
A versatile element, with gifts to trace,
Enchanting our world with its luminous pace.

THIRTY-FOUR

WITHOUT YOU

Magnesium, oh how you shine,
In the world of chemistry, you're divine.
You play a role in muscle movement,
Without you, our bodies would be stagnant.

Photosynthesis, you're a vital part,
Absorbing light and playing your part.
In Chlorophyll, your presence is felt,
Without you, plants wouldn't be able to help.

But there's more to you than just science,
In fireworks, you create a dazzling alliance.
Your bright light and sparks fill the night,
A sight that's truly a magnificent delight.

Magnesium, you're a versatile element,
A source of beauty and power, so evident.
From muscle movement to fireworks display,
Your presence is felt in every way.

THIRTY-FIVE

POWER SO POTENT

In the darkness of the earth's embrace,
Lies a shimmering gem, with radiant grace.
Magnesium, the element of fire,
Igniting passions with its burning desire.

In muscles strong, it dances and weaves,
Contracting and relaxing with every heartbeat it perceives.
From the twitch of a smile to a sprinting stride,
Magnesium's presence, an invisible guide.

In chlorophyll's embrace, it basks in the sun,
Nourishing plants, a battle it has won.
Photosynthesis, a magical dance,
Magnesium's touch, a life-giving chance.

In the night sky, a spectacle unfolds,
Colors bursting, as the story unfolds.

Fireworks explode, painting the air,
Magnesium's brilliance, beyond compare.
    With a blinding light, it takes flight,
Filling the world with pure delight.
A moment of awe, a memory to keep,
Magnesium's magic, forever deep.
    So let us marvel at this wondrous element,
A giver of life, a power so potent.
From muscles to plants, and skies up high,
Magnesium, the star in life's grand symphony.

THIRTY-SIX

VERSATILE WAYS

In nature's realm, a marvel shines,
A metal rare, that forever binds,
Magnesium, with brilliance true,
A tale of wonder, I'll share with you.
    In fields of green, where sunlight weaves,
Magnesium dances among the leaves,
Photosynthesis, its faithful guide,
To harness energy, far and wide.
    With chlorophyll in its embrace,
Magnesium grants life's vibrant grace,
From the depths of soil to the sky above,
It nurtures growth, with endless love.
    But beyond the realm of nature's embrace,
Magnesium paints the night with fiery grace,
In fireworks' burst, it takes its flight,
A symphony of colors, pure and bright.

With a sizzle and a crackle, it lights the sky,
Spinning and shimmering, as time goes by,
Magnesium's spark, a dazzling sight,
A moment of magic, in the dark of night.

Oh, Magnesium, your wonders unfold,
In muscle's strength and stories untold,
From the greenest fields to the starry night,
You shine with brilliance, ever so bright.

So let us marvel at your versatile ways,
In nourishing plants and lighting up our days,
Magnesium, you're a treasure, it's true,
A symphony of life, forever anew.

ABOUT THE AUTHOR

Walter the Educator is one of the pseudonyms for Walter Anderson. Formally educated in Chemistry, Business, and Education, he is an educator, an author, a diverse entrepreneur, and he is the son of a disabled war veteran. "Walter the Educator" shares his time between educating and creating. He holds interests and owns several creative projects that entertain, enlighten, enhance, and educate, hoping to inspire and motivate you.

Follow, find new works, and stay up to date
with Walter the Educator™
at WaltertheEducator.com

www.ingramcontent.com/pod-product-compliance
Lightning Source LLC
LaVergne TN
LVHW020134080526
838201LV00119B/3777